Aluminium, Steel and Sugar Industry in India

Siva Prasad Bose

Published by Joy Bose

This book is dedicated to all the people who are working in the
Aluminium and steel and sugar industries in India.

Contents

Other books by Siva Prasad Bose

Preface

India's industrial landscape is vast and diverse, yet few sectors touch every aspect of our economy and daily life as deeply as aluminium, steel, and sugar. These three industries form the backbone of India's infrastructure, agriculture, transportation, and energy systems.

Aluminium is used in a variety of applications and products ranging from bicycles and aluminium foil. Steel is widely used throughout the engineering and construction industry, and in manufacturing of a variety of items including cars, appliances and medical instruments. Sugar is used widely in the food industry. India has an important position in the world in all of these industries, both as a producing country as well as a consuming country.

Whether it is the lightweight resilience of aluminium in modern construction, the enduring strength of steel in railways and skyscrapers, or the agricultural and cultural legacy of sugar—each represents a unique thread in the tapestry of India's growth story.

This book is a simple, accessible, and comprehensive introduction to these three vital industries. It is meant for:

- Curious laypersons who want to understand how the materials around them are made,
- Students preparing for exams in geography, economics, engineering, or industrial studies,
- Professionals seeking a quick reference or foundational overview of these sectors.

We begin by exploring the natural resources—bauxite, iron ore, and sugarcane—that fuel these industries. We then walk through the key manufacturing processes that transform these raw materials into usable products. The book also profiles the major companies shaping these sectors in India, from public sector giants like SAIL and NALCO to private leaders like Hindalco, Tata Steel, and Balrampur Chini.

Alongside, we highlight government policies, sustainability efforts, and the future trends that will define India's industrial journey in the coming decades—from green aluminium and smart steel plants to ethanol-powered vehicles and Industry 4.0.

Acknowledgements

<hr>

Important Books/ References Consulted

- McGraw Hill Concise Encyclopedia of Science and Technology. Sybil P Parker, Eds. Mc Graw Hill, 1984.
- Encyclopaedia of Occupational Health and Safety. Volume 1. Third Revised Edition, 1989. ILO Geneva.
- Wikipedia en.wikipedia.org
- Birla Industrial and Technological Museum in Kolkata

We are thankful to Midjourney AI for the AI generated art produced using it, that was used for some of the illustrations in this book.

Chapter 1: Introduction to Aluminium

In this chapter we go through an introduction to Aluminium and its properties and applications.

1.1 What is Aluminium

Aluminium is a metal found in abundance on the earth. Chemically speaking, its symbol is Al, its atomic number is 26 and it is found in group 3 of the periodic table. Pure Aluminium is soft, but it can be moulded into alloys with other metals and elements to increase its strength. Aluminium alloys are light and strong. They can be readily formed by process of metalworking, easily joined or cast. Because of all these characteristics aluminium has become the world's most used non-ferrous metal.

Aluminium is not found alone in nature but rather in chemical compounds in minerals such as aluminium silicate (derived from alumina or Al2O3 and silicon dioxide). It is found in various plants and rocks. When these minerals go into solution, depending on the chemical conditions, aluminium can be precipitated out of solution, usually by electrolysis of minerals such as alumina.

1.2 Favourable properties of Aluminium

The properties of aluminium that result in its widespread use include the following:

- Aluminium is abundantly found in the earth's crust, in rocks and plants
- Aluminium is light in weight, being one third of the weight of steel

- Aluminium is malleable and also readily mouldable with other metals
- Aluminium is resistant to most forms of corrosion by chemicals and by water including sea water.
- Aluminium is non-combustible and non-pyrophoric, which is important in applications involving inflammable and explosive materials handling or exposure.
- Aluminium is non-toxic, which is a useful property for food containers
- Aluminium is non-ferromagnetic, which is useful in electrical and electronics components
- Aluminium is a good conductor of electricity and can dissipate heat rapidly. It displays excellent electrical and thermal conductivity. However, a few specific alloys have been developed with high degrees of electrical resistivity.
- Aluminium surfaces can be highly reflective. Radiant energy, visible light, radiant heat and electromagnetic waves are efficiently reflected, while anodized and dark anodized surfaces can be reflective absorbent. The reflectance of polished aluminium over a broad range of wavelengths, leads to its selection for a variety of decorative and functional uses.

1.3 What is Bauxite or Aluminium Ore

Bauxite is the ore for aluminium, being one of the principal raw materials for aluminium production. It is a name of a rock, composed of a variety of products comprising mainly of hydrated aluminium oxides and aluminous laterite. It is a mixture of minerals and not a mineral by itself.

Bauxite occurs in a variety of structures, textures and colours. The colour depends on the content of iron oxides and ranges from white to dark, red or brown.

Bauxite from mines is treated with concentrated sodium hydroxide, which results in a soluble solution of sodium aluminate. After filtration and heating with water, it gives aluminium hydroxide which is further heated to give aluminium.

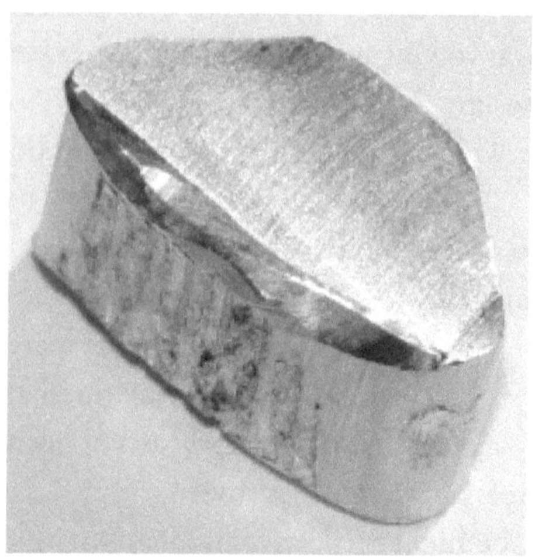

Figure: A chunk of Aluminium metal. By Unknown author - http://images-of-elements.com/aluminium.php, CC BY 3.0, https://commons.wikimedia.org/w/index.php?curid=9084427

Group / Period	1	2	3	4	5	6	7	8	9	10	11	12	13	14	15	16	17	18
1	1 H																	2 He
2	3 Li	4 Be											5 B	6 C	7 N	8 O	9 F	10 Ne
3	11 Na	12 Mg											13 Al	14 Si	15 P	16 S	17 Cl	18 Ar
4	19 K	20 Ca	21 Sc	22 Ti	23 V	24 Cr	25 Mn	26 Fe	27 Co	28 Ni	29 Cu	30 Zn	31 Ga	32 Ge	33 As	34 Se	35 Br	36 Kr
5	37 Rb	38 Sr	39 Y	40 Zr	41 Nb	42 Mo	43 Tc	44 Ru	45 Rh	46 Pd	47 Ag	48 Cd	49 In	50 Sn	51 Sb	52 Te	53 I	54 Xe
6	55 Cs	56 Ba	71 Lu	72 Hf	73 Ta	74 W	75 Re	76 Os	77 Ir	78 Pt	79 Au	80 Hg	81 Tl	82 Pb	83 Bi	84 Po	85 At	86 Rn
7	87 Fr	88 Ra	103 Lr	104 Rf	105 Db	106 Sg	107 Bh	108 Hs	109 Mt	110 Ds	111 Rg	112 Cn	113 Nh	114 Fl	115 Mc	116 Lv	117 Ts	118 Og

57 La	58 Ce	59 Pr	60 Nd	61 Pm	62 Sm	63 Eu	64 Gd	65 Tb	66 Dy	67 Ho	68 Er	69 Tm	70 Yb
89 Ac	90 Th	91 Pa	92 U	93 Np	94 Pu	95 Am	96 Cm	97 Bk	98 Cf	99 Es	100 Fm	101 Md	102 No

Figure: Periodic Table, Aluminium is symbol Al, row 3 column 13, atomic number 13. User:Double sharp, based on File:Simple Periodic Table Chart-en.svg by User:Offnfopt, CC BY-SA 4.0 <https://creativecommons.org/licenses/by-sa/4.0>, via Wikimedia Commons

1.4 Applications of aluminium

Aluminium is used in a variety of sectors, which include the following:

- Construction industry, such as doors, windows, and cladding
- Transportation, including the aerospace and shipbuilding industries
- For household utensils and food containers, because of its non-toxic nature
- Packaging of food, drink and medicines, such as aluminium foil and soda cans
- Because of its high conductivity and mechanical strength, aluminium is widely used in lay-live, high voltage aluminium steel covered reinforced transmission cables. It also has wide use for making overhead conductors, switchboards, coils and capacitors.
- Machinery and car components
- For water treatment and making a variety of medicines
- In the electrical and electronics industries, due to its non-magnetic properties.
- For making silver paints and mirrors
- In welding and metallurgy

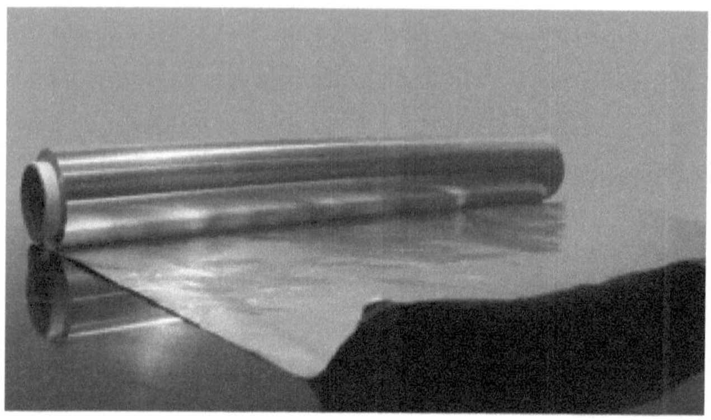

Figure: A roll of Aluminium foil. MdeVicente, CC0, via Wikimedia Commons

Figure: Aluminium can for soda. Marcos André, CC BY 2.0 <https://creativecommons.org/licenses/by/2.0>, via Wikimedia Commons

Figure: Aluminium products. Taken from Birla Industrial and Technological Museum in Kolkata, courtesy of Hindalco industries

Chapter 2: Aluminium Metallurgy

In this chapter we go through the process of Aluminium metallurgy.

2.1 What is Aluminium metallurgy

Aluminium metallurgy is the term for the refining of aluminium from its ore and its subsequent purification and processing, shaping it into usable products.

The most widely used technology for producing aluminium consists of two steps:

- **Bayer Process**: Extraction and purification of alumina (Al_2O_3) from ores such as Bauxite. The result is refined alumina.
- **Hall-Heroult process**: Smelting aluminium from alumina using electrolysis of the oxide after it has been dissolved in a fused cryolite bath.

Bauxite is the most used raw material for the production of aluminium. It is reddish in colour and usually found not too deep from the surface, so it is extracted from the ground by open cast mining.

2.2 Bayer Process for Alumina extraction from Bauxite

The aluminium extraction and refinement process from Bauxite, also called **Bayer process**, is as follows:

- Aluminium ore, primarily bauxite is crushed and ground.

- The ore is then digested at high temperatures and pressure. It is strongly heated at temperatures up to 1300 degrees Celsius in a strong solution of caustic soda (Sodium Hydroxide or NaOH).

- The gibbsite, which is a mineral form of aluminium hydroxide or alumina trihydrate or $Al(OH)_3$, along with bohemite or diaspore in the bauxite reacts with the caustic soda to form a soluble sodium aluminate. The residue, known as red mud, contains insoluble impurities. The red mud is separated from the solution.
- After cooling, the solution is seeded with regulated synthetic gibbsite and agitated.
- A large part of the alumina that is present in the solution then crystallizes out as gibbsite. Alumina contains Aluminium Oxide with about 0.3 to 0.8% soda, 0.1% iron oxide and silica, and trace amounts of other oxides.

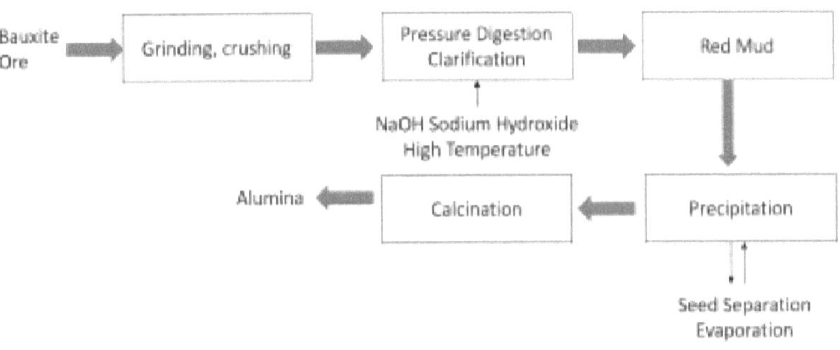

Figure: Flow of the refinement process (Bayer Process) for extraction of Alumina from Bauxite Ore

Figure: A Hall-Heroult Industrial cell. By Kashkhan at English Wikipedia, CC BY-SA 3.0, https://commons.wikimedia.org/w/index.php?curid=6864903

2.3 Hall-Heroult Process for extraction from Aluminium by Electrolysis

For aluminium metal production from the refined alumina by electrolysis, the **Hall-Heroult process** is as follows:

- In a smelter, alumina is dissolved in cells or vats, which are large insulated rectangular steel shells lined with carbon containing a molten electrolyte or bath consisting mostly of a sodium salt called cryolyte. The bath usually contains 2-8% alumina.
- Blocks of carbon anodes are hung from above the cells, with their lower ends extending to within 3.8 cm of the molten metal, which forms a layer under the molten bath. The heat needed to keep the belt molten is supplied by the electrical resistance of the bath as current passes through it.

- The passage of direct current through the electrolyte decomposes the dissolved alumina. The aluminium metal in liquid or molten form is deposited on the cathode, and oxygen is deposited on the anode that is gradually consumed.
- The smelting process is continuous. Alumina is added, anodes are replaced and molten aluminium is periodically siphoned off without interrupting the current to the cells.
- Molten aluminium is siphoned from the smelting cells into large crucibles.
- From there the aluminium metal is cast using various methods. It may be poured directly into moulds to produce foundry ingots (ingot casting), or transferred to holding furnaces for further refining or alloying with other metals to form fabricating ingots, through pig casting or continuous casting. As it comes from the cell, primary aluminium averages around 99.8% purity.

2.4 Extraction of aluminium from Scrap

In plants that are not adjacent to a smelter, it is necessary to remelt charges, usually consisting of mill scrap or returned scrap with enough primary metal to provide composition control. Scrap recycling has grown over the years to considerable economic importance and provides a valuable source of aluminium at a much lower energy cost than extraction of aluminium from primary sources.

Figure: Model of a pig casting machine. Taken from Birla Industrial and Technological Museum in Kolkata

2.5 Direct Chill Process for Casting of Aluminium Ignots

The most common type of casting methods for solid ingots of aluminium and other metals is called the DC or **Direct Chill process**.

The process involves the following steps:

- The molten aluminium metal is poured into a short mould, about 7.5 to 15 cm deep that is cooled by water. An outer layer of the metal solidifies in the mould.
- The base of the mould is a separate platform that is lowered gradually into a pit, while the metal is solidifying.
- The frozen outer shell of the metal retains the still liquid portions when the shell is past the mould wall.
- After leaving the mould, water is sprayed into the new aluminium ingot until it solidifies.

2.6 Conclusion

In this chapter, we have discussed about the process of producing aluminium from the bauxite ore. Aluminium metallurgy showcases the blend of chemistry and engineering that transforms raw bauxite into high-purity, high-value metal. With both primary and secondary production pathways, aluminium remains central to industrial progress.

Chapter 3: Aluminium Alloys

In this chapter we go through some aspects related to the common alloys of aluminium.

3.1 What are aluminium alloys

Aluminium alloys are substances formed by the addition of one or more elements, usually metals, to aluminium. The principal alloying element usually is one of magnesium, silicon, copper, zinc, tin, nickel or manganese. The alloying of aluminium produces a range of properties in the alloys, including enhanced strength, hardness, corrosion resistance, and electrical properties.

Aluminium alloys have wide industrial applications, including the following:

- **Aerospace**: Lightweight, high-strength parts

- **Automobiles and bicycles**: Frames and panels
- **Construction**: Extrusions for doors and windows
- **Consumer electronics**: Casings and internal components

3.2 Classification of Alloys

- **Wrought Alloys**: Mechanically worked; designated by a 4-digit code (e.g., 6061, 7075)
- **Cast Alloys**: Cast into shapes; used where complex shapes and lower cost are required

Each alloy is tailored to a specific use case based on physical and mechanical properties.

In wrought alloys, which constitute the greatest use of aluminium, the alloys are identified by four-digit numbers of the form nxxx when n denotes the alloy type and the principal alloying element. The international alloy designation system has a numbering process of 4 digits, with 1000 series meaning essentially pure aluminium.

Iron and silicon are commonly present as impurities in aluminium alloys, although the amount may be controlled to produce specific mechanical or physical properties. Also, minor amounts of other elements may be added to the alloys to achieve specific properties, such elements include Cr, Zr, V, Lead (Pb) and Bi. Titanium may be added to achieve a specific cast structure.

3.2 How are aluminium alloys fabricated

Aluminium alloys can be:

- Rolled into sheets
- Forged into structural parts
- Drawn into wires
- Cast into complex components

Joining methods include welding, riveting, and bolting, depending on application

In addition, pieces produced separately can be assembled using a common joining procedure to build up complex shapes and structures.

3.3 Conclusion

In this chapter we have discussed aluminium alloys and how they are fabricated. Aluminium alloys combine the lightness of aluminium with the strength and functionality required for a vast range of industries, making them indispensable in modern life.

Chapter 4: Aluminium Plants and Bauxite Mines in India

In this chapter, we discuss about the availability of aluminium ore in India in the form of Bauxite reserves.

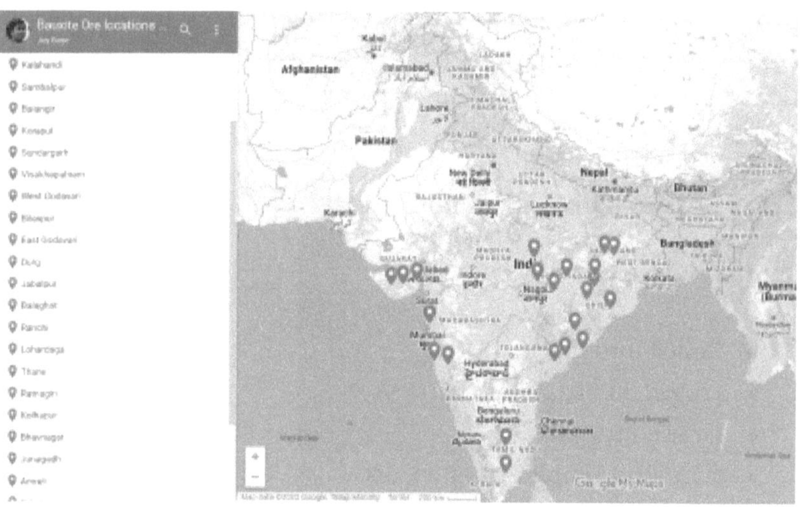

Figure: Map showing locations of Bauxite Mines in India. Made with Google My Map.

4.1 Bauxite reserves in India

India has huge reserves of aluminium ore or bauxite, for which it ranks 5th in the world. It has reserves of 2300 million tonnes or around 7% of the world reserves. India produced around 16.29 tons of Aluminium in 2010.

Bauxite is found mainly in the coastal, hill and peninsular areas of India.

The main bauxite producing states in India include Odisha, Gujarat, Jharkhand, Maharashtra, Chhattisgarh, Tamil Nadu, and Madhya Pradesh.

Of these, Odisha state is the largest producer of bauxite in India, producing about half of the country's total production. Kalahandi, Sambalpur, Balangir, Koraput and Sundargarh districts in Odisha belong to the bauxite producing belt.

Other important districts include the following:

- Vishakhapatnam and East and West Godavari districts in Andhra Pradesh
- Bilaspur and Durg in Chattisgarh
- Jabalpur and Balaghat in Madhya Pradesh
- Ranchi and Lohardaga in Jharkhand
- Thane, Ratnagiri and Kolhapur in Maharashtra
- Bhavnagar, Junagarh and Amreli in Gujarat
- Salem and Madurai in Tamil Nadu.

Figure: Bauxite mine in India. Taken from NALCO

4.2 Conclusion

In this chapter we have discussed the location of important Bauxite reserves in India.

Chapter 5: Aluminium Companies in India

In this chapter, we discuss the main applications of aluminium and the main aluminium producing companies in India.

5.1 Applications of aluminium in India

Due to its widespread availability and relatively low price as compared to other metals, aluminium is the most widely used non-ferrous metal in India.

Aluminium is widely used for making kitchen equipment such as pots and pans.

Figure: Anodised Aluminium pan. By FiveRings, Public domain, via Wikimedia Commons.

Aluminium is also used in a variety of industries due to its favourable characteristics such as lightweight, corrosion resistance and durability as discussed earlier, as well as the variety of uses and properties of its alloys.

Aluminium is widely used in India by the transport industry for making vehicles of all kinds. Aluminium is used in making the body of aircraft. It is used in making a variety of road vehicles such as bicycles, buses, trucks and vans. It is used in making railway coaches and goods wagons. It is similarly used in making ships and boats.

Aluminium is used widely in the electrical industry, such as for making power cables, capacitors, coil, switchboards etc. It is used by the manufacturing industry to make different kinds of machinery and equipment. It is used by the food industry for food packaging and aluminium foil. It is used in the construction industry for making doors and windows and cladding and weatherproof light weight constructions such as canopies and tents.

5.2 The main aluminium companies in India

Aluminium companies in India include the following:

- National Aluminium Company or NALCO
- Hindalco Industries
- Vedanta Group comprising of a number of smaller companies including Bharat Aluminium Company or BALCO

Together, these three groups dominate the aluminium production in India. In this chapter we discuss these three groups and other aluminium companies in more detail.

Figure: Aluminium smelter. Taken from NALCO

5.3 NALCO (National Aluminium Company)

From   Wikipedia   https://en.wikipedia.org/wiki/National_Aluminium_Company

National Aluminium Company Limited, abbreviated as NALCO, (incorporated in 1981) is a government company having integrated and diversified operations in mining, metal and power under the ownership of the Ministry of Mines and Government of India. Presently, the Government of India holds a 51.5% equity in NALCO.[2] It is one of the largest integrated bauxite–alumina–aluminium–power complex in the country, encompassing bauxite mining, alumina refining, aluminium smelting and casting, power generation, rail and port operations.[1] The company is the lowest-cost producer of metallurgical grade alumina in the world and lowest-cost producer of bauxite in the world as per a Wood McKenzie report. With sustained quality products, the company's export earnings accounted for about 42% of the sales turnover in the year 2018–19 and the company is rated as third-highest net export earning CPSE as per a Public Enterprise Survey report.

5.4 Hindalco Industries

From Wikipedia https://en.wikipedia.org/wiki/Hindalco_Industries

Hindalco Industries Limited an Indian aluminium and copper manufacturing company, is a subsidiary of the Aditya Birla Group.[4] Its headquarters are at Mumbai, Maharashtra, India. The company has annual sales of US$25 billion and employs around 20,000 people.[3] It is listed in the Forbes Global 2000 at 895th rank.[5] Its market capitalisation by the end of May 2013 was US$3.4 billion.[6] Hindalco is one of the world's largest aluminium rolling companies and one of the biggest producers of primary aluminium in Asia.[7]

5.5 Vedanta Group

From Wikipedia: https://en.wikipedia.org/wiki/Vedanta_Limited

Vedanta Limited is an Indian multinational mining company headquartered in Mumbai, India, with its main operations in iron ore, gold and aluminium mines in Goa, Karnataka, Rajasthan and Odisha.[4]

5.6 Other companies

Other Aluminium companies in India include:

- India Foils Limited
- Sacheta Metals
- Jindal Aluminium
- Madras Aluminium Company Limited (MALCO)
- Indian Aluminium Company Limited (INDAL)
- Century Extrusions Ltd.

Another major Aluminium company was Bharat Aluminium Company Limited (BALCO). It was sold to Vedanta in 2000 by the Government of India.

5.7 Conclusion

In this chapter, we have discussed some important uses of aluminium as well as the companies in India that produce it, both in the government sector as well as the private sector.

Chapter 6: Introduction to Steel

In this chapter we introduce steel, its alloys and its main applications.

6.1 What is Steel and what are Steel Alloys?

Steel is the name given to any of a great number of alloys that contains the great element iron (Fe) as the major component and small amounts of carbon (C) as the major alloy element. Unlike pure iron, which is soft and ductile, steel is strong, versatile, and can be hardened or tempered for different applications. It is one of the most important materials in modern civilization, found in everything from buildings and bridges to cars and kitchenware.

These alloys are more properly referred to as Carbon steels, and make up well over 90% of the tonnage of steels produced throughout the world. Small amounts, generally of the order of a few percent, of other elements such as manganese, silicon, chromium (Cr), molybdenum (Mo) and nickel (Ni) may also be present in carbon steels. However, when large amounts of alloying elements are added to iron to achieve special properties, other terms are used to describe the alloys. For example, to produce stainless steel huge amounts around 12% of more of chromium are added.

Steel is made from iron ore, which are naturally occurring compounds of iron with other elements such as oxygen to produce oxides. Iron ore is mined from the ground and transformed into steel using various processes. The important oxides in iron ore are magnetite, hematite, goethite, limonite and siderite. They vary in colours from greyish to deep red. Of these, magnetite (Fe_3O_4) and haematite (Fe_2O_3) are the major sources. These are found in abundance in India, especially in states like Odisha, Chhattisgarh, Jharkhand, and Karnataka. Iron is extracted through smelting processes, then converted into various types of steel through controlled mixing and heat treatment.

Figure: Hemetite, a form of Iron Ore. CC BY-SA 2.0 br, https://commons.wikimedia.org/w/index.php?curid=333366

6.2 Types of Steels according to Carbon Content

There are three kinds of steel, classified according to the carbon content:

- **Low-carbon steels** are commonly referred to as mild steels and contain less than 0.25% carbon. These steels are usually hot worked and are produced in large tonnages for beams and other structural applications.
- **Medium carbon steels** contain between 0.25 to 0.70% carbon and are frequently used for machine components that require high strength and good fatigue resistance.
- **High carbon steels** contain more than 0.7% carbon and are in a special category on account of their high hardness and low toughness. This combination of properties makes the high carbon steels good for applications where wear resistance is important and the comprehensive loading minimizes brittle fracture that might develop on tensile loading.
- When elements like chromium or nickel are added in

significant amounts, the steel becomes **alloy steel** or **stainless steel**, offering corrosion resistance and high temperature stability.

6.3 Applications of Steel

Steel is the foundation of modern infrastructure and industry:

- **Construction**: Skyscrapers, bridges, pipelines
- **Transport**: Automobiles, railways, ships
- **Machinery**: Gears, tools, industrial equipment
- **Consumer goods**: Appliances, cutlery, furniture

Its strength, recyclability, and adaptability make steel an industrial workhorse.

6.4 Conclusion

In this chapter we have introduced steel and its alloys, as well as the various types of steel. Steel has transformed modern society by enabling safe, large-scale construction, transportation, and industrial production. Its adaptability and strength have made it one of the most widely used materials in the world.

Chapter 7: Steel Manufacturing Process

In this chapter we introduce aspects related to the metallurgical processes for the manufacture of steel.

7.1 What is Steel Manufacturing Process

Steel manufacturing is the name given to the sequence through which blast furnace iron, scrap iron and alloying additions are processed to combine and to separate them into desired compositions of steel and residual slag.

Steel manufacturing involves converting raw iron ore and scrap into different grades of steel using various chemical, thermal, and mechanical techniques. The process not only involves purification but also careful control over composition and structure.

The manufacture of steel is a tremendous industry with extensive ramifications resulting from the complex interrelation of technical and economic considerations.

7.2 Types of Steel Manufacturing Process

Chemically, all steel making processes may be classified into acidic or basic, depending on the refractory and slag combination.

- **Acid processes** use silica (SiO_2) refractories throughout and are able to accommodate slags that become saturated with this component under operating conditions. The acid systems can be used to eliminate C, Mn and Si from the charge, but require select raw materials within final steel specifications for parts.
- **Basic processes** use Magnesite (MgO) or equivalent refractions

in the portions of the furnace that contact molten slag and metal and accommodate lower silical slags with compensating amounts of lime (CaO). These systems can eliminate C, Mn and Si as effectively as can the acidic systems. They also eliminate P and appreciable amounts of S.

Figure: A steel mill. AI generated art by Midjourney AI.

Technologically, processes can be grouped into the following three main types, in order to use available raw materials and heat sources effectively for particular grades of steel:

- **Pneumatic processes**: In such processes. all heat is derived from the initial heat content of the charge materials, principally molten, and the thermochemical balance of the refining sections. The selective oxidation of the refining is accomplished by blowing air or commercial oxygen. The Basic Oxygen Furnace (BOF) uses molten pig iron and scrap, blows high-pressure oxygen to oxidize carbon and impurities and produces

bulk steel efficiently
- **Open-hearth processes (now nearly obsolete)**: It is used to melt pig iron and scrap over several hours. In such processes, the major source of heat is the combustion of fuel, usually gas or oil. This in turn depends for its success on the regenerative principle of preheating air to attain steel making temperatures efficiently. Blast furnaces or open oxygen furnaces are used for the purpose.
- **Electric Arc Furnace Processes**: Here the major source of heat is electric current, which can be of the type arc, or resistance or both. Because this heat can be produced in the presence or absence of oxygen, electric furnaces can operate in a neutral or non-oxidizing atmosphere or vacuum, and thus are either preferred or required for alloys with significant amounts of easily oxidized elements and for other special grades where an improved control of gases is required. The process melts scrap steel using electric arcs, is flexible and energy-efficient, suitable for high-quality and specialty steels and is ideal for a circular economy as it recycles scrap

Figure: An open hearth furnace in Ukraine. By Viktor Mácha - Own work, CC BY-SA 4.0, https://commons.wikimedia.org/w/index.php?curid=42224671

Figure: Model of an electric arc furnace. Taken from Birla Industrial and Technological Museum in Kolkata.

Figure: Steel smelting. AI generated art by Midjourney AI.

7.3 Heat Treatment or Metallurgy

Metallurgy is the name given to a procedure of heating and cooling a material without melting. Plastic deformation may be included in the sequence of heating and cooling steps, thus defining a thermo-mechanical treatment.

Heat treatment modifies the microstructure of steel to enhance specific properties:

- **Annealing**: Softens steel for improved ductility
- **Hardening**: Increases strength and wear resistance
- **Tempering**: Reduces brittleness after hardening
- **Surface Hardening**: Focused treatment on surface layer (carburizing, nitriding, flame hardening)

Other objectives of heat treatment include improved formality, improved machinability, and improved Dimensional Stability.

Heat treatments are often characterized by special names such as annealing, normalizing, stress relief anneals, process anneals, hardening tempering, mar tempering, interstitial annealing, carburizing, nitriding, solution anneal, aging, precipitation, hardening and thermomechanical treatment.

All metals and alloys in common use are heat treated at some stage during processing. Iron alloys, however, respond to heat treatments in a unique way because of the multitude of phase changes which can be induced, and it is thus convenient to discuss heat treatment of ferrous and non-ferrous metals separately.

7.4 Structural Steel

Structural steel is the term used for steel used in engineering structures, usually manufactured by either the open-hearth or electric furnace process. The exception is carbon steel plates and shaped whose thickness is 11 mm or less and which are used in structure subject to static loads only. These products may be made from acid-Bessemer steel. The physical properties and chemical composition are governed by standard specifications. Structural steel can be fabricated into different shapes for various construction purposes.

7.5 Stainless Steel

Stainless Steel is the generic name commonly used for the entire group of iron-based alloys which exhibit phenomenal resistance to rusting and corrosion because of chromium content. The amount of chromium exceeding 10.5% with carbon amount kept suitably low make the iron rust-proof.

Other alloy elements such as nickel (Ni) and molybdenum (Mo) can also be added to the base composition to produce variety as well as improved properties.

Over 100 different stainless steels are produced commercially, out of which half are standardized grades. Some of these can be more properly termed as stainless irons since they do not harden as steels. Others are true steels, to which corrosion resistance becomes an added feature. Still others are not properly either steel or iron and can be considered as new classes of materials.

Other types of speciality steels include:

- **Tool Steel**: Used for cutting tools and molds
- **High-Strength Low-Alloy (HSLA)**: High strength with good formability

7.6 Surface hardening of steel

Surface hardening is the term used for the selective hardening of the surface layer of a steel product by one of several processes which involve changes in microstructure with or without changes in composition. Surface hardening imparts a combination of properties to the finished product not produced by bulk heat treatment alone. Among these are high wear resistance good toughness on impact, better resistance to failure by fatigue resulting from cyclic loading, and resistance to surface indentation by localized loads. The use of surface hardening is often favoured in order to get greater flexibility in manufacturing.

The principal surface hardening processes include the following:

- **Carburizing**: This introduces carbon into the surface layer of low-carbon steel parts and converts that layer into high-carbon steel, which can be quench hardened by appropriate heat treatment.

- The modified carburizing processes of **carbonitriding, cyaniding and liquid carburizing**: These processes, in addition to supplying carbon, introduce nitrogen into the surface layer. This has a beneficial effect on subsequent heat treatment.
- **Nitriding**: In this process, only nitrogen is supplied and reacts with special alloy elements present in the steel.
- **Flame hardening and induction hardening:** These depend on a heat treatment applied selectively to the surface layer of a medium carbon steel.
- **Surface working**: This is performed by shot peening, surface rolling, or presenting improved fatigue resistance by producing a stronger case, compressive stresses, and a smoother surface.

7.7 Ferroalloys

Ferroalloys are members of an important group of metallic raw materials required for the steel industry. They are mixtures of iron and other elements (like manganese, silicon, or chromium) added to steel to improve strength, resistance to wear, and heat tolerance. They are essential in producing alloy and stainless steels.

They are the principal source of such additions as silicon and manganese which are needed for even the simplest plain-carbon steels, and chromium, vanadium, tungsten, titanium and molybdenum which are used in both low and high alloy steels.

Ferroalloys are unique in that they are brittle and otherwise unsuited for any service application, but they are important as the most economical source of these elements for use in the manufacturing of the engineering alloys. These same elements can be obtained at a much higher cost as essentially pure metals. The ferroalloys contain significant amounts of iron and usually have a lower melting range than the pure metals and are therefore dissolved by the molten steel more readily than the pure metal. In other cases, the other elements in the ferroalloy serve to protect the critical element against oxidation during solution and thereby give higher recoveries. Ferroalloys are used both as deoxidizers and as a specified addition to give particular properties to the steel.

7.8 Conclusion

In this chapter we have discussed various processes for manufacturing of steel. Steel manufacturing is a highly technical and evolving process. With innovations in recycling and energy efficiency, the industry is continuously adapting to meet global demand and sustainability goals.

Chapter 8: Iron and Steel Plants and Companies in India

In this chapter we introduce the iron and steel industry in India. We briefly discuss the main steel plants and companies in India.

8.1 Iron Ore locations in India

Steel is extracted from iron ore. Hence, the places with a higher amount of iron ore will also be able to support steel production.

Iron ores are minerals found in the soil from which iron is extracted, which can later be refined to produce steels.

India is richly endowed with iron ore, ranking in the top iron ore producers globally. The total recoverable reserves of iron ore in India are about 9602 MT or million tonnes of hematite and 3408 MT of magnetite.

The states in India with major production of iron ore are as follows:

- Chhattisgarh: Bastar, Dantewada
- Madhya Pradesh
- Karnataka: Bellary, Chitradurga
- Jharkhand: Singhbhum, Ranchi
- Odisha: Keonjhar, Sundargarh
- Goa
- Maharashtra
- Andhra Pradesh
- Kerala
- Rajasthan
- Tamil Nadu

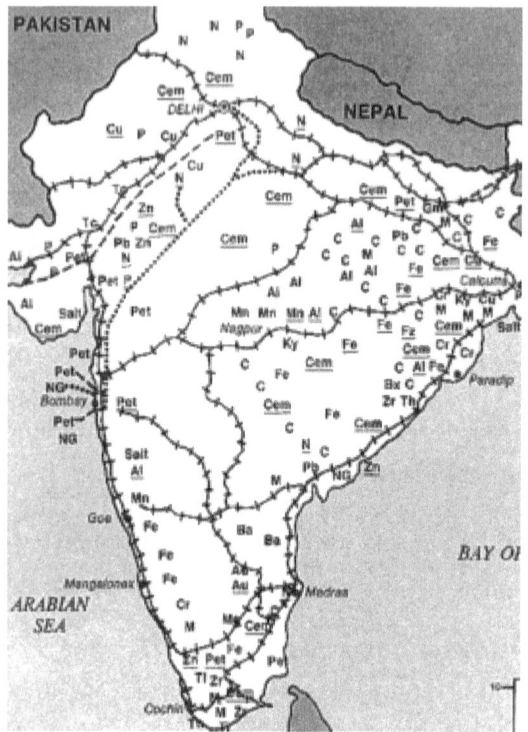

Figure: Distribution of iron (Fe) and other minerals in India, according to the United States Geological Survey. United States Geological Survey, Public domain, via Wikimedia Commons

8.2 Introduction to Steel Industry in India

India is one of the world's biggest producers of steel.

As per data from the government of India (steel.gov.in) India produced 118.2 MT or million tonnes of Crude steel in 2021, making it the second largest producer of crude steel in the world. India is also the second largest consumer of finished steel, with 106.23 MT in 2021. The industry includes both public and private sector giants.

8.3 Steel plants and companies in India

The steel companies in India are a mixture of public sector (PSUs) and private sector. Some of the top steel companies are as follows:

- **Steel Authority of India Limited or SAIL**. It is the largest steel producing company. It is a PSU owned by the government and operates huge integrated steel plants in Rourkela, Bhilai, Bokaro, Asansol as well as special steel plants in Bhadrawati, Durgapur and Salem. The Bhilai steel plant in Chattisgarh is the flagship plant of SAIL and one of India's biggest steel plants.
- **Tata Steel Limited**: It is one of the oldest and most respected private steel companies in India, part of the Tata conglomerate and is the second largest steel company after SAIL. It has steel plants in Jamshedpur.
- **JSW Steel**: it was formed by the merger of Ispat Steel and Jindal Vijaynagar steel. It is India's second largest steel company, being part of the JSW group.
- **Jindal Steel and Power Limited** and Jindal Iron and Steel: They are privately owned, and work in multiple industries.
- **Rashtriya Ispat Nigam Limited**: It is a government owned PSU
- **Essar Steel / ArcelorMittal**: This is privately owned and has plants in multiple countries. It is a major player in the global steel industry.

While PSUs like SAIL laid the foundation of India's steel sector post-independence, private giants have driven modernization, global expansion, and innovation. Together, they contribute to employment, exports, and national infrastructure.

8.4 Conclusion

In this chapter we have briefly looked at the iron ore producing areas in India and the biggest steel making companies and steel plants. India's steel sector is vital for economic growth, infrastructure, and global trade. With robust reserves, policy support, and a growing domestic market, it is poised for even greater expansion in the coming years.

Chapter 9: Introduction to Sugar

In this chapter we introduce the sugar industry. We start with the sugarcane plant and the areas of sugarcane cultivation.

Sugar, a staple of the Indian diet and a symbol of festivity, is also a major industry that impacts millions of lives—from rural farmers to global traders. Derived mainly from sugarcane in India, sugar production supports agricultural livelihoods, fuels ethanol-based energy, and contributes to exports.

9.1 Introduction to Sugarcane

Sugarcane plant is the source of sugar. Its Latin name is *Saccarum Officianarum* which is a member of the grass family. The sugarcane crop originated in New Guinea around 15000 to 8000 BC and was later moved by primitive peoples westward into Southeast Asia and India and eastward into Polynesia. Most current commercial varieties of sugarcane are hybrids.

It thrives in tropical and subtropical climates and is known for its high sucrose content. In India, it is cultivated widely in states such as Uttar Pradesh, Maharashtra, Karnataka, Tamil Nadu, and Bihar.

Each stalk of sugarcane can grow up to 6 meters in height and is composed of fibrous tissue filled with sweet juice. This juice is extracted and processed to make various types of sugar.

Figure: Cut Sugarcane. By Rufino Uribe - caña de azúcar, CC BY-SA 2.0, https://commons.wikimedia.org/w/index.php?curid=1155927

9.2 Sugarcane Cultivation

Sugarcane is an exceptionally tall perennial grass that is cultivated, usually in plantations, in tropical and sub-tropical regions, for its sucrose content and by-products such as molasses and bagasse. The plant grows in clumps of cylindrical stalks measuring 1.25-7.25 cm in diameter and reaching 6-7 m in height. These stalks, containing a sap from which sugar is processed, are segmented by joints, each of which contains a bud or eye which will sprout when planted.

Figure: A sugarcane field. AI generated art by Midjourney AI

9.3 Areas of Sugarcane cultivation

Sugarcane is grown throughout the Caribbean, in central and south America and in India, the pacific islands, Australia, central and south Africa, Mauritius and the southern United States. In some of the smaller countries, the sugar crop forms a large proportion of export wealth, such as up to 80% in Mauritius.

The main use of sugarcane cultivation is the production of sugar. However, it may instead be fermented and distilled to produce rum. The cellulose material, bagasse, which remains after pressing, may be used in the production of paper.

Under favourable conditions and appropriate use of pesticides and fertilizers, the cane grows rapidly. To ensure the maximum sugar content, up to 10-17% of the total weight, it must be harvested immediately after it reaches its last growth period. In hand-cutting using long knives such as machetes, the cane is cut down near the ground, the leaves are stripped off and the stalks are chopped into convenient short lengths ready for processing. Hand cutting prevails where there is abundant labour or where the peculiarities of the cane fields present obstacles to machinery. Large numbers of migrant or seasonal workers are employed during harvesting.

9.4 Other Sources of Sugar

While sugarcane is the primary source of sugar in India, globally sugar is also produced from sugar beet, corn (as high-fructose corn syrup), and honey. However, these are either limited or niche in the Indian context.

9.5 Conclusion

In this chapter we have discussed some basics of sugar from sugarcane plant and about cultivation of sugarcane. Sugar is not just a product but a cultural and economic force in India. Understanding its origin and cultivation is the first step toward appreciating the scale of the industry.

Chapter 10: Process for Sugar Production

In this chapter we discuss the process for production of sugar starting from the cultivation of plants till the production of sugar, both from sugarcane plant and beet plant. Corn syrup and honey are other sources of sugar.

Sugar manufacturing is a multi-step process that begins in the fields and ends in crystalline sweetness. The steps differ slightly depending on the source (cane or beet), but the principles remain similar.

10.1 Process for separation of sugar from sugarcane

Sugar is generally removed from sugarcane in large factories by either the milling or the diffusion process:

- **Harvesting**: Sugarcane is harvested manually or mechanically and transported to mills within 24–48 hours to prevent sucrose loss.
- **Milling process:** This crushes the sugarcane stalks as they pass between a series of large metal rollers separating the fibre or bagasse from the juice that is laden with sugar.
- **Diffusion:** This process separates sugar from finely cut stalks by dissolving it in hot water or hot juice and then filtering.
- **Clarification**: Juice is treated with lime to remove non-sugar impurities and filtered.
- **Evaporation**: Water is removed from the clarified juice using vacuum evaporators.
- Concentrated juice is seeded with sugar crystals and cooled to promote crystal formation. The process by which sugar crystals are formed is known as **crystallization**.

- **Centrifugation**: The mixture is spun in high-speed centrifuges to separate crystals from molasses.

Bagasse and molasses are the principal by-products from processing sugarcane. Molasses consists of syrup that contains up to 55% sugar along with other impurities.

The uses of the by products is as follows:

- **Bagasse**: Burned in boilers to generate electricity
- **Pressmud**: Fertilizer or soil conditioner
- **Molasses**: Used for ethanol, animal feed, or liquor production

10.2 Processing of Beet-Sugar

Apart from sugarcane, another source of sugar is the sugar beet plant, species name beta vulgaris. Sugar beet is more popular in Europe and North America. It is sliced, soaked in hot water to extract juice, and follows a similar refining process. While not widely grown in India, beet sugar is studied as a possible supplement.

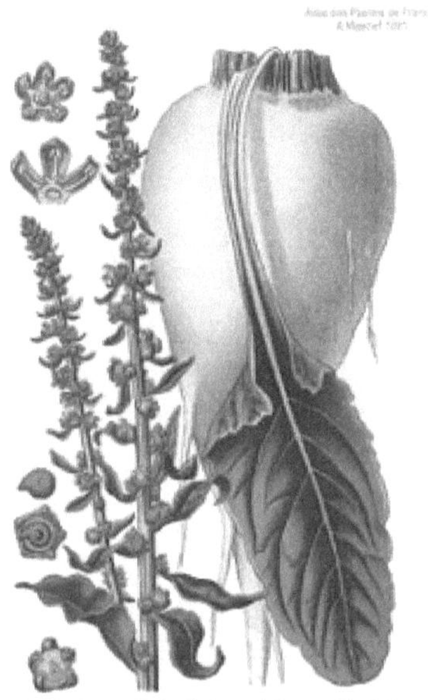

Figure: Sugar beet, illustration of root, leaf, and flowering patterns. Amédée Masclef, Public domain, via Wikimedia Commons

The process of producing sugar from beet consists of many steps which have been refined and improved over time.

The beets have sugar content between 15% to 18%. They are first cleaned in a beet washer. Then they are cut in beet slicers and the cossettes thus obtained are conveyed via a scalder into the diffuser where most of the sugar contained in the beets is extracted by exhaustion in hot water. The de-sugarized cossettes, now called pulps, are pressed mechanically and dried mostly by heat. The pulps still contain many nutrients and are used as animal feed.

The raw juice obtained in the diffuser contains, in addition to sugar, many non-sugar impurities which are precipitated by adding lime and carbon dioxide and then filtered off. The raw juice then has a sugar content of 12 to 14%. This thin juice is then concentrated in evaporators to about 65 to 70% dry matter. This thick juice is boiled in a vacuum pan at a temperature of 70 degrees Centigrade until crystals form. This massecuite is discharged into mixers and then the liquid surrounding the crystals is spun off.

The low syrup thus separated from the sugar crystals still contains sugar that can be crystallized. The desugaring process is continued until it is no longer economical. The syrup left after the last crystallization is called molasses.

After drying and cooling, the sugar is mostly stored in silos, where it can be kept for a practically unlimited period of time, if sufficiently air conditioned. But processing makes it possible to obtain white sugar of high quality in line with international standards.

The molasses slices contain about 60% of sugar and together with the non-sugar impurities constitute valuable animal feed as week as an ideal culture medium for many microorganisms. Part of the molasses is added to the sugar exhausted cossettes (pulps) before they are dried, making animal feed. Molasses are also used for the production of yeast and alcohol.

10.3 Working conditions in beet sugar factories

In highly mechanised beet sugar industry, the beet is transformed into sugar during a so-called campaign of 3 to 4 months. During this period the factories work continuously, i.e., personnel work in 3 to 4 shifts. Their number is slightly increased by auxiliary campaign personnel in order to cope with the workload. During the off season, the necessary repair, maintenance and expansion work is done in the factories. This is why the sugar factories require a large number of highly skilled workers.

The working conditions in beet sugar factories is not very different from the rest of the industry. Trade unions are usually present that conclude collective agreements with employers. Legislation in India to protect workers is also quite exhaustive.

10.4 Hazards and their prevention in beet sugar factories

Beet sugar factories are mostly situated in rural areas i.e. in towns with 1000 inhabitants or less. Good hospitals and other infrastructure like schools may be difficult to find in such conditions.

In the working conditions of beet sugar factories, there are no toxic gases or airborne dusts. The temperatures are high in some areas, around 38 to 40 degrees Celsius, but the relative humidity rarely exceeds 70%. In many cases, air conditioned electric control panels have been installed.

10.5 Production of Cane Sugar

In the sugar mill, the cane is crushed and the juice extracted by heavy rollers. For every 100 kg of cane treated, the yield is 75 kg of juice: an opaque, greenish-grey, alkaline liquid. The juice contains saccharose, glucose, laevulose, organic salts and acids in solution and is mixed with bagasse fibres, grit, clay, colouring matter, albumen and pectin in suspension. Because of the albumen and pectin, the juice cannot be filtered cold: heat and chemical agents are required to eliminate the impurities and to obtain saccharose.

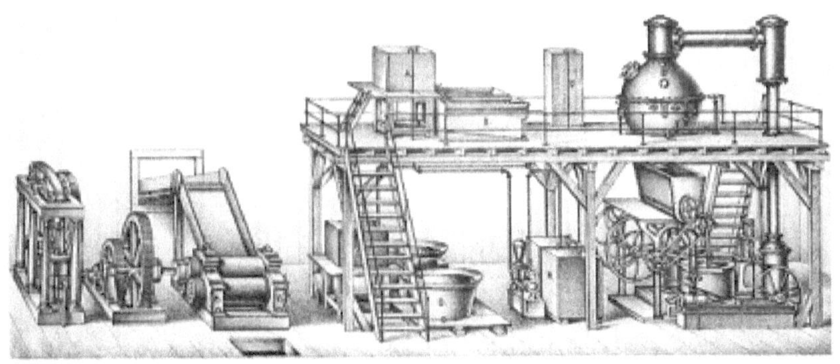

Figure: Sugarcane mill and boiling apparatus. Morris, Tasker & Co., Public domain, via Wikimedia Commons.

Clarification is obtained through heating and the addition of lime-based precipitants. Once clarified, the juice is concentrated by vacuum evaporation until it precipitates in the form of greyish crystals.

The concentrated juice or molasses contains 45% of water. Centrifuge treatment produces granulated sugar of a greyish hue, often called brown sugar, for which there is a market. White sugar, on the other hand, is obtained through a refining process. In the refineries, the brown sugar is dissolved with chemicals such as sulphuric anhydride and phosphoric acid, and is filtered with or without bone black, according to the purity desired.

The filtered syrup evaporates in vacuo until it crystallizes. Centrifugation is then applied until a white crystalline powder is obtained.

10.6 Honey as a source of sugar

Honey is another natural source of sugar. It is made from nectar or honeydew collected flowers by honeybees. Honeybees are cultivated in artificial hives or bee colonies for the purpose of honey extraction. Flower nectar has a high water content of up to 80%.

Figure: A jar of honey with a honey dipper. Scott Bauer, USDA ARS, Public domain, via Wikimedia Commons

Forager honeybees collect nectar from flowers and store in their honey stomachs, then return to the hive and regurgitate and transfer the nectar to the hive bees, which store, digest and regurgitate honey in their own honey stomachs. Part of the water gets evaporated in the hive, increasing the honey concentration, till it reaches the finished concentration of around 18% water content.

The honey is then extracted from the artificial hives. Smoke is used by the beekeepers to pacify the bees in the hives and enable safe extraction of honey.

Pure natural honey has good preservation properties and a long shelf life. Sometimes other sugars and syrups are added to honey before being sold.

10.7 Corn syrup as a source of sugar

Corn syrup is made from the starch of corn or maize. It contains varying amount of glucose and other oligosaccharides.

Corn syrup was manufactured historically by adding dilute hydrochloric acid to corn starch. The mixture is then heated under pressure, in what is known as the Kirchhoff process. Another manufacturing process is by adding an enzyme α-amylase to a mixture of corn starch and water, which breaks down the starch into oligosaccharides. Another enzyme called glucoamylase is then then added which further breaks the resulting compound into glucose molecules. Finally, the result is passed through a column loaded with D-xylose isomerase enzyme that transforms it into fructose.

Corn syrup is commonly used in food as a thickener and sweetener.

10.8 Conclusion

In this chapter, we have briefly discussed the process of manufacturing of sugar from various sources including sugarcane, sugar beet, honey and corn. From raw cane to refined crystals, sugar production is a marvel of agricultural and chemical engineering. Its by-products make it a zero-waste industry in many respects, contributing to food, fuel, and farm cycles.

Chapter 11: Sugar Industry in India

In this chapter we discuss the sugar industry in India.

11.1 Introduction to sugar industry of India

Sugarcane is a plant native to the Indian subcontinent. Sugar has been produced and used as a food flavouring in India since ancient times.

India is the world's largest consumer and second-largest producer of sugar. The industry supports over 50 million farmers and thousands of mill workers. With over 500 mills, it is both decentralized and vital to rural economies.

Most of the sugar from India is produced from sugarcane. Uttar Pradesh produces about 47% of sugar in India and is the highest sugar producing state, followed by Maharashtra (22%) and Karnataka. Maharashtra is known for its cooperative sugar mills.

11.2 India Sugar Mills Association

India Sugar Mills Association or ISMA is the main sugar organization in India. It is the industry body representing sugar mills. It was established in 1932 and includes more than 532 sugar mills across India in its members, including both government and private sugar mills. This includes 50% of the total sugar produced in India.

Figure: Sugar mill in India

11.3 Indian Institute of Sugarcane Research (IISR) and Central Sugarcane Research Institute (CSRI) and National Sugar Institute (NSI)

These are important government research institutes in India for research into various aspects of sugarcane production and sugar industry. IISR is located in Lucknow, Uttar Pradesh, CSRI in Coimbatore in Tamil Nadu and NSI in Kanpur, Uttar Pradesh. IISR and CSRI are under the Indian Council of Agriculture Research. NSI Kanpur is under the Department of Food and Public Distribution of the Government of India, and conducts technical research and training.

11.4 Role in Energy and Industry

Sugar mills are evolving into **biorefineries**, producing ethanol, bioelectricity, and organic fertilizers. Ethanol production, in particular, supports India's fuel blending goals and reduces oil imports.

11.5 Challenges Faced

- **Pricing conflict** between cane farmers and mill owners

- **Delayed payments** to farmers due to surplus and market volatility
- **Weather dependence**: Cane is water-intensive and sensitive to climate

11.6 Conclusion

The sugar industry is at the crossroads of tradition and transformation. With smart reforms and sustainable practices, it has the potential to lead India's green energy and rural prosperity revolutions.

Chapter 12: Government Policies and Industrial Reforms

India's aluminium, steel, and sugar industries are not shaped by natural resources and production technologies alone. They are also the result of deliberate planning, regulations, subsidies, and reform programs enacted by the government. In an increasingly competitive and environmentally conscious global market, India's industrial policy must balance domestic growth, global competitiveness, and sustainability. This chapter explores the key reforms, policy instruments, and regulatory frameworks that influence these three sectors.

12.1 Aluminium Policy Landscape

India has vast reserves of bauxite and a growing domestic demand for aluminium in infrastructure, aviation, power transmission, and packaging. Government policies are aimed at leveraging these advantages while reducing import dependence:

- **Mining Reforms**: Amendments to the Mines and Minerals (Development and Regulation) Act (MMDR) allow for private-sector participation in bauxite mining through auctions, enhancing transparency and efficiency.
- **PLI Scheme Inclusion**: The aluminium sector is lobbying for inclusion under the Production-Linked Incentive (PLI) scheme to boost domestic manufacturing.
- **Tariff Protection**: Import duties ranging from 7.5% to 10% help shield Indian producers from global price shocks and dumping.
- **Environmental Compliance**: Rules on fly ash reuse and red

mud disposal increase environmental accountability but add to compliance costs.

12.2 Steel Industry Policy Initiatives

The steel industry is critical to India's growth, particularly in infrastructure and manufacturing. Policy support has been both strategic and protective:

- **National Steel Policy 2017**: Aims to increase crude steel capacity to 300 million tonnes by 2030 and improve per capita steel consumption.
- **Make in India Drive:** Boosts demand via construction, housing, and defence procurement.
- **Steel Scrap Recycling Policy**: Promotes circular economy and reduces dependence on imported scrap.
- **Anti-Dumping Duties**: Imposed on imports from countries like China to protect domestic manufacturers from unfair pricing.

12.3 Sugar Sector Regulations

The sugar industry is intricately linked with agriculture and rural livelihoods, making its regulation both economically and politically sensitive:

- **Fair and Remunerative Price (FRP):** A statutory minimum price that mills must pay to farmers, ensuring income stability.
- **Ethanol Blending Policy:** Encourages mills to divert surplus sugar to ethanol, helping both the energy and agriculture sectors.
- **Export Subsidies:** Occasionally provided to clear excess stock and stabilize domestic prices.

- **SAP vs FRP**: Some states offer a higher State Advised Price (SAP), putting pressure on mill economics and creating centre-state friction.

12.4 Key Government Schemes by Industry

The table below summarizes the key government schemes and policies related to aluminium, steel and sugar industries.

Industry	Key Policies/ Schemes	Objectives
Aluminium	Import Duties, PLI Proposal, Mining Auction	Boost local production, reduce import reliance
Steel	National Steel Policy 2017, Scrap Policy	Build capacity, ensure fair trade, Achieve 300 MT capacity, circular economy
Sugar	FRP/SAP, Ethanol Blending, Export Subsidies	Support farmers, manage surplus, green energy

12.5 Integrated Policy Framework: Example of Sugar Industry

The sugar sector showcases a complex ecosystem of policies involving multiple stakeholders:

Farmers → Sell Cane → Sugar Mills → Process Sugar/Ethanol

↓ ↓

Govt Fixes FRP Govt Provides Export Subsidies

↓

Ethanol Blending Program (10%-20%)

This integrated approach enables flexibility, supports biofuel goals, and addresses overproduction challenges.

12.6 Challenges and Way Forward

Despite ambitious policies, several challenges remain:

- Balancing farmer interests with mill viability in sugar
- Managing environmental regulations without stifling ease of business
- Reducing import dependency while enhancing export competitiveness

The future lies in a coherent policy mix that harmonizes growth, equity, and sustainability.

Chapter 13: Sustainability and Environmental Challenges

Heavy industries, by their very nature, are resource-intensive. From open-cast mining in aluminium to water-intensive sugarcane cultivation and the high carbon emissions of steel, these sectors carry significant environmental burdens. In this chapter, we delve into the specific ecological challenges and emerging green solutions across these industries.

13.1 Sector-wise Environmental Footprint

Industry	Major Environmental Concerns	Sustainable Solutions
Aluminium	Red mud waste, high electricity usage, deforestation	Red mud bricks, renewable energy integration
Steel	CO2 emissions, particulate matter, slag disposal	Carbon capture, electric arc furnaces
Sugar	Water overuse, fertilizer runoff, effluents	Drip irrigation, ethanol production, organic methods

13.2 Environmental Regulations and Best Practices

- **Pollution Control Board (PCB) Norms**: Enforced across all states for emissions and effluent discharge.
- **Zero Liquid Discharge (ZLD)**: Now mandatory in many sugar mills to prevent groundwater contamination.
- **Reuse Mandates**: Fly ash in steel and red mud in aluminium must be reused for bricks or cement.
- **Afforestation Compensations**: Especially in bauxite mining,

Odisha has led with afforestation offsets.

13.3 Corporate Sustainability Leadership

- **NALCO:** Uses captive solar power and promotes red mud-based construction materials.
- **JSW Steel:** Among the first to adopt low-emission electric arc furnace technology in India.
- **Balrampur Chini Mills:** Implements bagasse-based power generation and ethanol production.

13.4 The Road Ahead

Future sustainability hinges on three pillars:

- **Technology**: Carbon capture, smart sensors, and AI for energy efficiency.
- **Policy Support**: Incentives for clean-tech investments.
- **Stakeholder Collaboration**: Government, industry, and civil society must act in concert.

Environmental sustainability is no longer an option but a necessity for long-term viability.

Chapter 14: Future Trends and Industry Outlook

As India aspires to become a global manufacturing hub, the future of aluminium, steel, and sugar will be shaped by megatrends like decarbonization, digitalization, and changing consumption patterns. This chapter outlines the trends that will define these industries in the coming decades.

14.1 Key Trends

- **Urbanization & Infrastructure Growth**: Driving demand for steel and aluminium in housing, transport, and smart cities.
- **Electric Vehicles (EVs)**: Increasing aluminium usage in lightweight battery casings and wiring.
- **Energy Transition**: Sugar mills evolving into integrated biorefineries producing ethanol and green electricity.
- **Industry 4.0**: Use of AI, IoT, and digital twins to optimize production and reduce waste.

Trend	Aluminium	Steel	Sugar
Green Manufacturing	Solar-powered smelters	Electric arc furnaces	Bioethanol expansion
Industry 4.0	Smart rolling mills	AI quality inspection	Drone-based crop analysis
Circular Economy	Scrap recycling	Scrap-to-steel plants	Molasses reuse
Export Opportunity	High-purity ingots	Structural steel	Sugar & ethanol exports

14.2 Sectoral Opportunities

Industry	Future Opportunities
Aluminium	Solar-powered smelters, green aluminium exports
Steel	Scrap-based production, carbon-neutral steelmaking (Green Steel)
Sugar	Export of ethanol, organic farming, decentralized bioenergy plants

14.3 Strategic Challenges

- Chinese Dominance: Competing with China's scale and pricing remains a hurdle.
- Climate Pressures: Meeting India's net-zero targets requires cleaner production.
- Labour Upskilling: Industry 4.0 requires a digitally savvy workforce.

14.4 India's Future Roadmap

2025: 15% ethanol blending target

2026: India becomes net exporter of aluminium

2028: Majority of steel from electric arc furnaces

2030: India reaches 300 MT steel capacity

2030: Net-zero sugarcane processing in top 5 states

14.5 Conclusion

The future of these industries lies in how well they adapt to sustainability imperatives, leverage new technologies, and respond to shifting global demands. India has the resources, talent, and ambition to lead—what remains is to execute with vision.

Chapter 15: Conclusion

India's aluminium, steel, and sugar industries are more than just economic sectors—they are lifelines of national development, employment, and global relevance. Through this book, we have explored their raw material origins, manufacturing processes, major companies, regional hubs, government policies, sustainability challenges, and future outlooks.

15.1 Why These Three Industries Matter

Each of these industries plays a foundational role in India's economy:

- **Aluminium** powers our infrastructure, packaging, and mobility with a lightweight and recyclable metal.
- **Steel** forms the backbone of buildings, machines, vehicles, and bridges—supporting a $3 trillion economy and beyond.
- **Sugar**, derived from one of India's most ancient crops, sustains livelihoods for millions of farmers while also fuelling a modern ethanol economy.

Together, they impact agriculture, industry, energy, exports, and employment in direct and indirect ways.

15.2 Key Takeaways from the Book

- **Resource Abundance**: India has large reserves of bauxite and iron ore, as well as vast sugarcane-producing regions. Harnessing these efficiently can make India a global leader.
- **Process Innovation**: From the Bayer and Hall-Héroult processes in aluminium, to electric arc furnaces in steel, to

advanced crystallization in sugar refining, each industry is a testament to scientific ingenuity.

- **Industrial Giants**: Companies like NALCO, Hindalco, Tata Steel, JSW, and Balrampur Chini have not only shaped domestic markets but are now making global inroads.
- **Government Role**: Reforms, pricing policies, and environmental regulation play a crucial role—both in enabling growth and addressing social or ecological externalities.
- **Sustainability Is Non-Negotiable**: All three sectors face pressure to go green—whether through recycling aluminium, decarbonizing steel, or reducing water usage in sugar.
- **The Road Ahead Is Technological**: AI, automation, smart manufacturing, and clean energy will be the growth engines of the future.

This book is designed as a friendly, accessible guide for anyone curious about Indian industry—from students preparing for exams, to working professionals, to general readers seeking to understand how industrial India works.

In a rapidly changing world, industries that adapt will thrive. India stands at a pivotal point in its industrial journey. By blending innovation, sustainability, and inclusive growth, these three industries can continue to shape not just the nation's economy—but also its future identity on the world stage.

We hope this book has sparked your interest in the aluminium, steel, and sugar industries of India and inspired you to explore further.

About the Author

Siva Prasad Bose is a writer of introductory guidebooks on different aspects of Indian laws. He is also a retired electrical engineer, retired after many years of service in Uttar Pradesh Power Corporation Limited. He received his engineering degree from Jadavpur University, Kolkata and has a law degree from Meerut University, Meerut and a Bachelor of Science from MMH University Ghaziabad. His interests lie in the fields of family law, civil law, law of contracts, and areas of law related to power electricity related issues.

Other books by Siva Prasad Bose

Introduction to Wills and Probate

Delays in Court Cases in India·

Introduction to negotiable instruments

Introduction to marriage laws in India

Neighbor Problems in India and what to do about them

Managing Court Cases with Mental Strength

Introduction to Patents and Patent Law in India

Introduction to Property Law in India

Don't miss out!

Visit the website below and you can sign up to receive emails whenever Siva Prasad Bose publishes a new book. There's no charge and no obligation.

https://books2read.com/r/B-A-RVCN-HUUBC

BOOKS 2 READ

Connecting independent readers to independent writers.

Did you love *Aluminium, Steel and Sugar Industry in India*? Then you should read *Introduction to Electricity Supply and Regulation in India*[1] by Siva Prasad Bose!

[2]

Electricity powers every aspect of modern India, yet few people truly understand how it is generated, regulated, and priced. This book changes that.Introduction to Electricity Supply and Regulation in India is a comprehensive yet accessible guide to both the technical and legal dimensions of India's power sector. Written for students, professionals, and curious readers alike, it breaks down complex subjects into clear, readable chapters, no engineering or law degree required.The book is organised into two parts:PART ONE: THE TECHNICAL SIDELearn how electricity is generated from coal, hydro, nuclear, solar, and wind sources. Understand how it travels from power plants through

---

1. https://books2read.com/u/ba6Rk6

2. https://books2read.com/u/ba6Rk6

transmission lines and substations to reach your home or factory. Chapters cover electric power generation, transmission systems, distribution networks, energy measurement, protective devices, and smart grid technology — giving you a solid grounding in how India's electrical infrastructure actually works.PART TWO: THE LEGAL AND REGULATORY FRAMEWORKTrace the history of electricity legislation in India from the colonial era to the landmark Electricity Act 2003. Understand the roles of CERC, SERCs, DISCOMs, and the Central Electricity Authority. Explore how tariffs are set, why electricity bills vary so widely across states, how privatisation of distribution companies is reshaping the sector, and what India's renewable energy targets mean for the future.This updated edition also covers recent developments including India crossing 500 GW installed capacity, the Revamped Distribution Sector Scheme (RDSS), the PM Surya Ghar rooftop solar scheme, Battery Energy Storage Systems, and the proposed Electricity Amendment Bill 2022.Whether you are a law student, an energy professional, a policy researcher, or simply a consumer who wants to understand their electricity bill, this book is your essential guide to one of India's most vital and rapidly changing sectors.

Read more at https://sivaprasadbosc.wordpress.com/.

About the Author

Siva Prasad Bose is an electrical engineer by profession. He is currently retired after many years of service in Uttar Pradesh Power Corporation Limited. He received his engineering degree from Jadavpur University, Kolkata and has a law degree from Meerut University, Meerut. His interests lie in the fields of family law, civil law, law of contracts, and any areas of law related to power electricity related issues.

Read more at https://sivaprasadbose.wordpress.com/.

www.ingramcontent.com/pod-product-compliance
Lightning Source LLC
Chambersburg PA
CBHW051805130726
47987CB00003B/1119